algebraforall yellow level

elizabeth warren PhD

About the Author

Elizabeth Warren has been involved in Mathematics Education for more than 30 years. During this time she has actively worked in both the university and school levels and engaged both elementary and secondary schoolteachers in professional development activities. She is presently conducting research into Patterns and Algebra.

Algebra for All, *Yellow Level*

Copyright 2006 ORIGO Education
Author: Elizabeth Warren, PhD

Warren, Elizabeth.
Algebra for All: yellow level.
ISBN 1 921023 01 5.
1. Algebra - Problems, exercises, etc. - Juvenile literature. I. Title.
512

For more information, contact
North America
Tel. 1-888-ORIGO-01 or 1-888-674-4601
Fax 1-888-674-4604
sales@origomath.com
www.origomath.com

Australasia
For more information,
email info@origo.com.au
or visit www.origo.com.au for other contact details.

All rights reserved. Unless specifically stated, no part of this publication may be reproduced, or copied into, or stored in a retrieval system, or transmitted in any form or by any means, electronic, mechanical, photocopying, recording, or otherwise, without the prior written permission of ORIGO Education. Permission is hereby given to reproduce the blackline masters in this publication in complete pages, with the copyright notice intact, for purposes of classroom use authorized by ORIGO Education. No resale of this material is permitted.

ISBN: 1 921023 01 5

10 9 8 7 6 5 4 3 2 1

INTRODUCTION 2

Equivalence and Equations

Birds in Trees 6
Balancing Act 8
Same Again 10
Picture This 12
More and Less 14
Beads and Boxes 16

Patterns and Functions

Shape Maker 18
Pretty Patterns 20
Growing Bigger 22
Missing Pieces 24
Missing Parts 26
Great Grids 28
Odd and Even 30
Growing Numbers 32
Nifty Numbers 34
Fruit Rules 36
Helping Hampers 38
Guess the Rule 40

Properties

Turnarounds 42
Make a Match 44
Fish Bowls 46
Order Option 48
Paper Strips 50
Take a Turn 52
No Change 54

Representations

Frogs in a Pond 56
Mix and Match 58
Bits in Boxes 60
Reading Pictures 62
Peanuts for Grabs 64

ANSWERS 66

INTRODUCTION

What is algebra?

Algebraic thinking commences as soon as students identify consistent change and begin to make generalizations. Their first generalizations relate to real-world experiences. For example, a child may notice a relationship between her age and the age of her older brother. In the example below, Ali has noted that her brother Brent is always 2 years older than her.

Ali's age	Brent's age
8	10
9	11
10	12
11	13

Over time these generalizations extend to more abstract situations involving symbolic notation that includes numbers. The above relationship can be generalized using the following symbolic notation.

Ali + 2 = Brent   A + 2 = B

Algebraic thinking uses different symbolic representations, such as unknowns and variables, with numbers to explore, model, and solve problems that relate to change and describe generalizations. The symbol system used to describe generalizations is formally known as algebra. Following the example above, Ali wonders how old she will be when Brent is 21 years old. We can solve a problem such as this by "backtracking" the generalization (A = 21 – 2).

Why algebra?

Identifying patterns and making generalizations are fundamental to all mathematics, so it is essential that students engage in activities involving algebra. Many practical uses for algebra lie hidden under the surface of an increasingly electronic world, such as specific rules used to determine telephone charges, track bank accounts and generate statements, describe data represented in graphs, and encrypt messages to make the Internet secure. Algebraic thinking is more overt when we create rules for spreadsheets or simply use addition to solve a subtraction problem.

Algebra involves the generalizations that are made regarding the relationships between variables in the symbol system of mathematics.

What are the "big ideas"?

The lessons in the *Algebra for All* series aim to develop the "big ideas" of early algebra while supporting thinking, reasoning, and working mathematically. These ideas of equivalence and equations, patterns and functions, properties, and representations are inherent in all modern curricula and are summarized in the following paragraphs.

Equivalence and Equations

The most important ideas about equivalence and equations that students need to understand are:

- "Equals" indicates equivalent sets rather than a place to write an answer
- Simple real-world problems with unknowns can be represented as equations
- Equations remain true if a consistent change occurs to each side (the balance strategy)
- Unknowns can be found by using the balance strategy.

Patterns and Functions

This idea focuses on mathematics as "change". Change occurs when one or more operation is used. For example, the price of an item bought on the Internet changes when a freight charge is added. It is important for students to understand that:

- Operations almost always change an original number to a new number
- Simple real-world problems with variables can be represented as "change situations"
- "Backtracking" reverses a change and can be used to solve unknowns.

Properties

Students will discover a variety of arithmetic properties as they explore number, such as:

- The commutative law and the associative law exist for addition and multiplication but not for subtraction and division
- Addition and subtraction are inverse operations, as are multiplication and division
- Adding or subtracting zero and multiplying or dividing by 1 leaves the original number unchanged
- In certain circumstances, multiplication and division distribute over addition and subtraction.

Representations

Different representations deepen our understanding of real-world problems and help us identify trends and find solutions. This idea focuses on creating and interpreting a variety of representations to solve real-world problems. The main representations that are developed in this series include graphs, tables of values, drawings, equations, and everyday language.

INTRODUCTION

About the series

Each of the six *Algebra for All* books features 4 chapters that focus separately on the "big ideas" of early algebra — Equivalence and Equations, Patterns and Functions, Properties, and Representations. Each chapter provides a carefully structured sequence of lessons. This sequence extends across the series so that students have the opportunity to develop their understanding of algebra over a number of years.

About the lessons

Each lesson is described over 2 pages. The left-hand page describes the lesson itself, including the aim of the lesson, materials that are required, clear step-by-step instructions, and a reflection. These notes also provide specific questions that teachers can ask students, and subsequent examples of student responses. The right-hand page supplies a reproducible blackline master to accompany the lesson. The answers for all blackline masters can be found on pages 66-73.

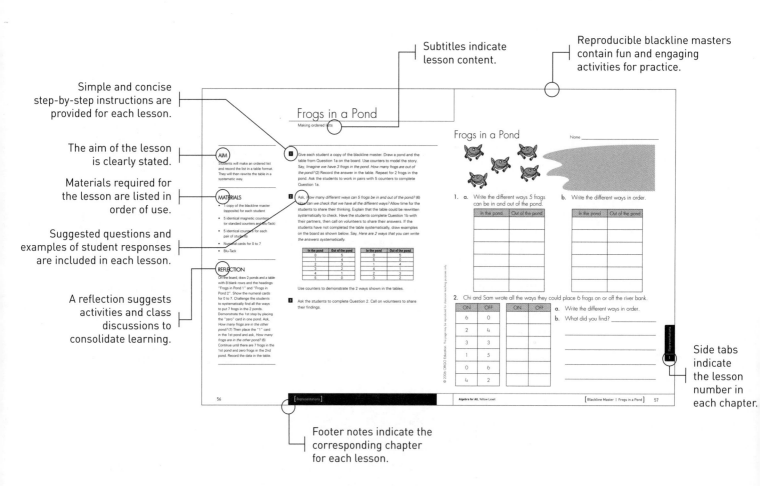

Assessment

Students' thinking is often best gauged by the conversations that occur during classroom discussions. Listen to your students and make notes about their thinking. You may decide to use the rubric below to assess students' mathematical proficiency in the tasks for each lesson. Study the criteria, then assess and record each student's understanding on a copy of the Assessment Summary provided on page 74. Although the summary lists every lesson in this book, it is not necessary to assess students for all lessons.

A	The student fully accomplishes the purpose of the task. Full understanding of the central mathematical ideas is demonstrated. The student is able to communicate his/her thinking and reasoning.
B	The student substantially accomplishes the purpose of the task. An essential understanding of the central mathematical ideas is demonstrated. The student is generally able to communicate his/her thinking and reasoning.
C	The student partially accomplishes the purpose of the task. A partial or limited understanding of the central mathematical ideas is demonstrated and/or the student is unable to communicate his/her thinking and reasoning.
D	The student is not able to accomplish the purpose of the task. Little or no understanding of the central mathematical ideas is demonstrated and/or the student's communication of his/her thinking and reasoning is vague or incomplete.

Birds in Trees

Exploring equivalence involving addition

AIM

Students will use real-world addition situations to model equivalence and then write equations involving addition.

MATERIALS

- Red, green, and blue magnetic counters (or standard counters and Blu-Tack)
- 1 copy of the blackline master (opposite) for each student

REFLECTION

Draw 2 trees on the board. Model the story with counters. Say, *One tree has 5 blue birds, 2 green birds, and 3 red birds, and the other has 2 blue birds, 3 green birds, and 5 red birds. Is there the same number of birds in each tree? How do we write this?* Invite several volunteers to share their suggestions, then write **5 + 2 + 3 = 2 + 3 + 5** on the board. Point to the equals symbol and ask, *What is this telling us?* (Each side is the same.)

1. Draw 2 different-sized trees on the board. Use counters to model the story. Say, *There are 2 trees in this garden. There are 4 red birds and 3 green birds in the large tree, and 5 red birds and 2 green birds in the small tree. How many birds are in the large tree?* (7) *How many birds are in the small tree?* (7) Ask volunteers to write the number sentences below each tree. Then ask, *Is there the same number of birds in each tree?* (Yes.) *How can we show this?* Elicit several responses then write **4 + 3 = 5 + 2** on the board.

2. Write **5 + 2 = 1 + 6** on the board and say, *Tell me a story that describes this number sentence.* For example, "There are 2 trees in a garden; one has 5 red birds and 2 green birds in it, and the other has 1 red bird and 6 green birds. There is the same number of birds in each tree." Repeat for **6 + 4 = 7 + 3** and then **5 + 2 + 3 = 8 + 2**. Discuss the equal sets on each side of the "equals" symbol.

3. Complete Question 1 on the blackline master with the class. Then read Question 2 with the students. Make sure they understand that there must be the same number of flowers in each tree. After the students have completed their drawings, they should ask a friend to write a number sentence to match their pictures.

[Equivalence and Equations]

Birds in Trees

Name _____

1. Write a number sentence to match each picture.

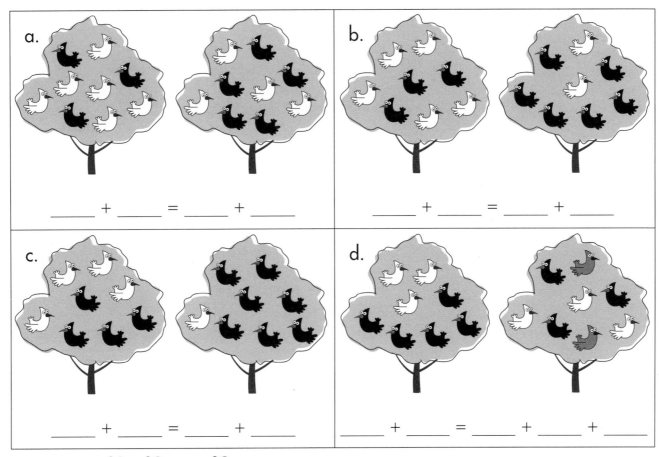

2. Draw some 🌼, ✼, and ❋ in each tree. Make the total number of flowers in each tree the same.

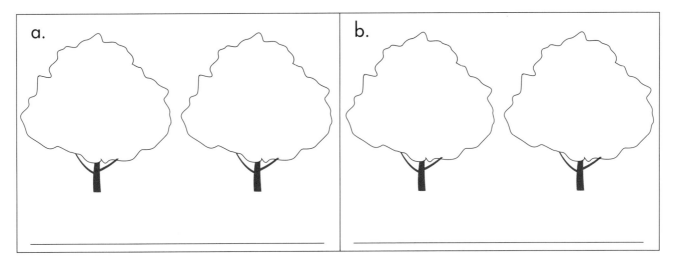

Algebra for All, Yellow Level [Blackline Master | Birds in Trees] 7

Balancing Act

Using balance to represent addition equations

AIM

Students will use balance scales to represent equations such as 2 + 3 = 5. In doing so, they will see that "equals" means "balances", and begin to explore the symmetrical property of the equals symbol.

MATERIALS

- 1 set of balance scales
- Red, green, and blue large bear counters (5 of each)
- Sign showing the equals symbol (=)
- Blu-Tack
- 1 copy of the blackline master (opposite) for each student

REFLECTION

Ask, *Is 8 = 6 + 1 true or not true?* Invite volunteers to share their thinking. Then ask, *How do we know if a number sentence is true?* (The sets on each side of the equals symbol are the same amount.) *If we flip a true number sentence, will it still be true?*

TEACHING NOTE

The symmetrical property of the equals symbol is shown by flipping a true equation; for example, 4 + 1 = 3 + 2 is the same as 3 + 2 = 4 + 1. This property is true for equality regardless of the operation.

1 Place 5 red bear counters on the right side of the balance scales and say, *There are 5 red bears on one side of these scales. How can we balance the scales?* Call on volunteers to share their solutions. Then place 3 blue and 2 green bear counters on the left side of the scales as shown on the right.

Ask, *Do 3 blue bears and 2 green bears weigh the same as 5 red bears? Are the scales balanced? How many bears are on each side? What symbol do we use to show that the scales are balanced?* Stick the "=" sign to the front of the scales and then ask, *How can we write this?* Record **3 + 2 = 5** on the board. Say, *3 add 2 equals 5.* Turn the scales around so that the 5 red bears are on the left, as shown, and ask, *How can we write this?* Record **5 = 3 + 2** on the board and say the new equation. Repeat for **4 + 1 = 5**. Discuss the symmetrical property of the equals symbol (see Teaching Note).

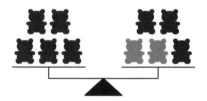

2 Read the blackline master with the students. Allow time for them to complete the questions. For each question, point out that the scale in the bottom row should show the flip (reverse) of the scale in the top row. Call on volunteers to share their answers to Question 2 with the class.

Balancing Act

Name _____

1. Draw counters to show different ways to balance the number 5. Then, flip each set of scales and write number sentences to match.

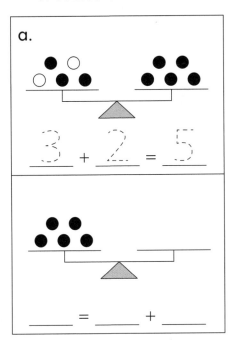

a. 3 + 2 = 5

___ = ___ + ___

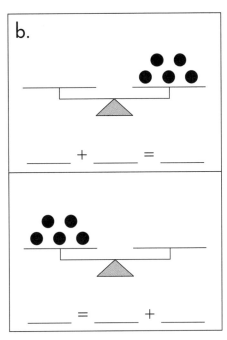

b. ___ + ___ = ___

___ = ___ + ___

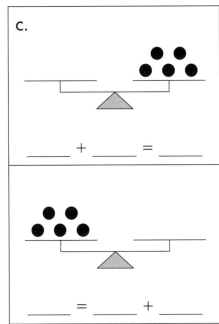

c. ___ + ___ = ___

___ = ___ + ___

2. Draw counters to show different ways to balance a number greater than 6. Then, flip each set of scales and write number sentences to match.

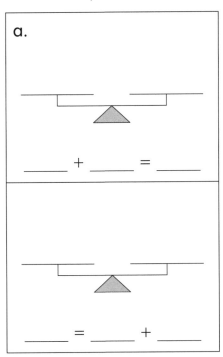

a. ___ + ___ = ___

___ = ___ + ___

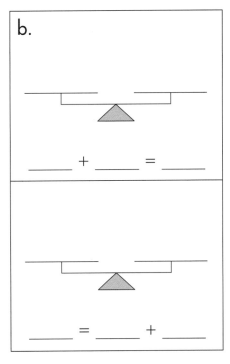

b. ___ + ___ = ___

___ = ___ + ___

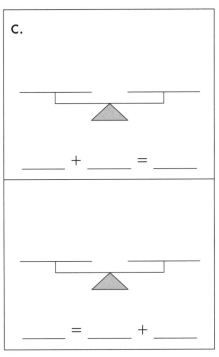

c. ___ + ___ = ___

___ = ___ + ___

Same Again

Using balance to represent addition equations

AIM

Students will use balance scales to represent equations such as 6 + 3 = 5 + 4. They will see that equivalent collections have the same value, and further explore the symmetrical property of the equals symbol.

MATERIALS

- 1 set of balance scales
- Red and green large bear counters (9 of each)
- Sign showing the equals symbol (=)
- Blu-Tack
- 1 copy of the blackline master (opposite) for each student

REFLECTION

Ask, *Is 5 + 3 = 6 + 1 true or not true?* Invite volunteers to share their thinking and ask, *How do we know if a number sentence is true?* (The value on each side of the equals symbol is the same.) Ask, *If we flip a true number sentence, will it still be true?*

1 Place 5 red and 2 green bear counters on one side of the scales and say, *There are 7 bears on one side of the scales. How many ways can we balance the scales?* Call on volunteers to share their solutions. Then place 3 red and 4 green bear counters on the other side of the scales, as shown below.

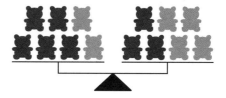

Ask, *Do 3 red bears and 4 green bears weigh the same as 5 red bears and 2 green bears?* (Yes.) *How can we write this?* Elicit several suggestions. Record **3 + 4 = 5 + 2** on the board and stick the "=" sign to the front of the scales. Turn the scales around to show 5 + 2 = 3 + 4 and ask, *How can we write this?* Repeat for **1 + 6 = 5 + 2**.

2 Place 6 red and 3 green bear counters on one side of the scales and ask, *How many different ways can we balance the scales?* As the students offer suggestions, model their answers on the scales and write the matching equations on the board. Ask, *How else can we write these equations?* Flip each equation and write the new number sentence on the board. Discuss the symmetrical property of the equals symbol.

3 Read the blackline master with the class. Allow time for the students to complete the questions. For each question, point out that the scale in the bottom row should show the flip (reverse) of the scale in the top row. Call on volunteers to share their answers to Question 2.

[Equivalence and Equations]

Same Again

Name _____

1. Draw counters to show different ways to balance the number 7. Then, flip each set of scales and write number sentences to match.

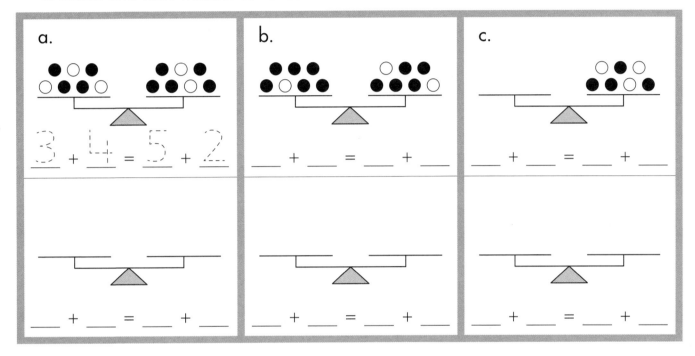

2. Draw counters to show different ways to balance a number greater than 7. Then, flip each set of scales and write number sentences to match.

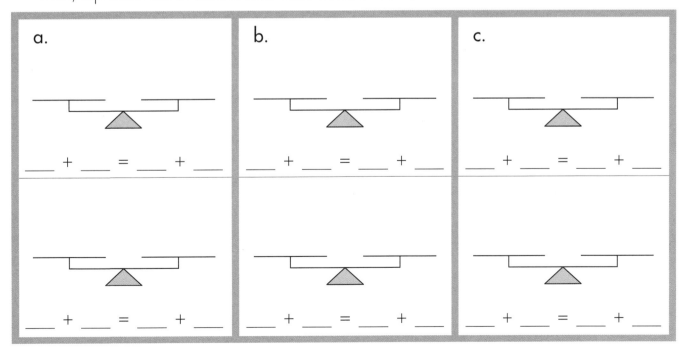

Picture This

Using balance to represent "equal" and "not equal"

AIM

Students will use balance scales to explore number sentences that are equal and not equal.

MATERIALS

- 1 set of balance scales
- Red and blue large bear counters (10 of each)
- 2 signs: equals symbol (=) and "not equal" symbol (≠)
- Blu-Tack
- Picture of a "parking allowed" or other similar sign
- 1 copy of the blackline master (opposite) for each student

REFLECTION

Have the students write "equal" and "not equal" number sentences that they can quickly figure out mentally. For example, increasing and decreasing the two numbers by the same amount (13 + 6 = 11 + 8), or increasing one of the numbers by 1 (12 + 8 ≠ 12 + 9).

1 Place 5 red and 3 blue bear counters on the left side of the scales, and 4 red and 4 blue bear counters on the right. Ask, *How many bears are on the left? How many bears are on the right? Do we have the same number on each side? How do we write this?* Call on volunteers to share their suggestions, then record **5 + 3 = 4 + 4** on the board. Add 1 red bear to the left side then ask, *Do we still have the same amount on each side? How do we write this?* On the board, write **6 + 3 ≠ 4 + 4**. Repeat for **4 + 2 = 5 + 1**.

2 Discuss the language "equals" and "not equal", and use the signs to model. Show the students a picture of a "parking allowed" sign and ask, *If we see this on the side of a street, what does it mean?* (Parking is allowed.) *How do we make this into a "no parking allowed" sign?* (We draw a diagonal line through it.) Relate the "≠" sign to the "no parking allowed" street sign.

3 Ask, *How do we know if a number sentence is equal or not equal?* (If there is the same amount on each side, it is equal; if there is a different amount on each side, it is not equal.) *Is this always true?* Elicit several responses. Write **12 + 3 ____ 11 + 4** on the board and ask, *Is this equal or not equal? How do you know?* Continue the discussion for **11 + 7 ____ 14 + 4**, **21 + 13 ____ 21 + 14**, and **36 + 5 ____ 37 + 4**. Ask questions such as, *Did you count on or back? Did you look at the differences in the numbers?*

4 Read and complete Question 1a on the blackline master together with the students. Make sure they understand that they must draw 6 counters on each side to make it balance. Allow time for the students to complete Question 1. Question 2 is more open-ended. Read the instruction and after sufficient time, call upon volunteers to share their answers.

[Equivalence and Equations]

Picture This

Name _____

1. Complete the missing parts.

a. $6 = 2 + 4$ 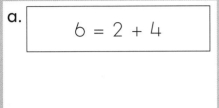	b. 	c. $3 + 4 = 6 + 1$ 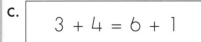
d. $3 + 1 \neq 4 + 2$ 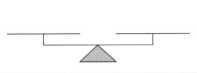	e. 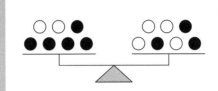	f. $6 + 2 \neq 7 + 3$ 
g. $3 + 5 \neq 4 + 2$ 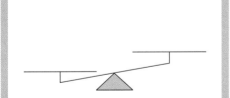	h. 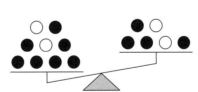	i. $7 + 4 \neq 9 + 1$ 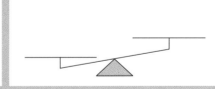

2. Write a number sentence. Draw counters to match.

a. 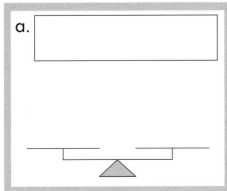	b. 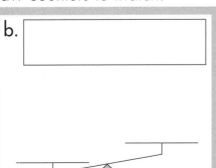	c. 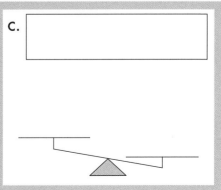

Algebra for All, Yellow Level [Blackline Master | Picture This]

More and Less

Using balance to represent "greater than" and "less than"

AIM

Students will use balance scales to determine whether one side of a number sentence is "greater than" or "less than" the other side, for example, 6 + 7 is greater than 4 + 5.

MATERIALS

- 1 set of balance scales
- Yellow and purple large bear counters (11 of each)
- 4 signs: "not equal" symbol (≠), "greater than", "more than", and "less than"
- Blu-Tack
- 1 copy of the blackline master (opposite) for each student

REFLECTION

Ask, *Can we make "greater than" and "less than" number sentences using more than 2 numbers on either side?* Call on volunteers to offer suggestions. Write the number sentences on the board and ask, *Which side is "greater than"? How do you know? Which side is "less than"? How do you know?*

1 Place 5 purple and 3 yellow bear counters on the right side of the scales, and 6 purple and 4 yellow bear counters on the left. Point to the left and ask, *How many purple and yellow bears on this side?* (6 + 4) *How many purple and yellow bears are on the other side?* (5 + 3) *Are there more bears on one side than on the other side?* (Yes.) *Which side has more bears?* (The left side.) *What are some different ways we can say this?* Use the signs to model the answers: 6 add 4 is not equal to 5 add 3; 6 add 4 is more than 5 add 3; and 6 add 4 is greater than 5 add 3. Turn the scales around and ask, *How do we say this?* (5 add 3 is less than 6 add 4.)

Repeat for 7 + 2 and 6 + 5. In the discussion, focus on the language used to describe the inequalities "more than", "greater than", and "less than".

2 Read the blackline master together with the students. Allow time for them to complete the questions. Call on volunteers to share their answers.

[Equivalence and Equations]

More and Less

Name _____

1. Place a ✓ on the side that has more.

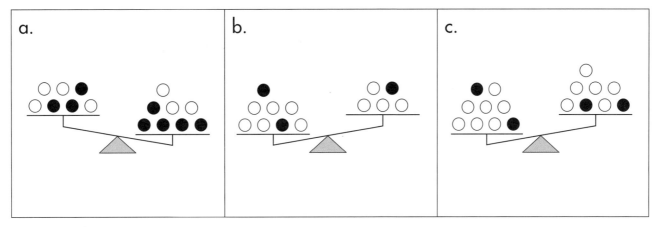

2. Place a ✓ on the side that has less.

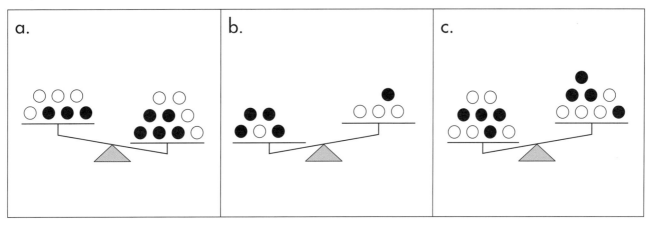

3. Draw counters to complete these.

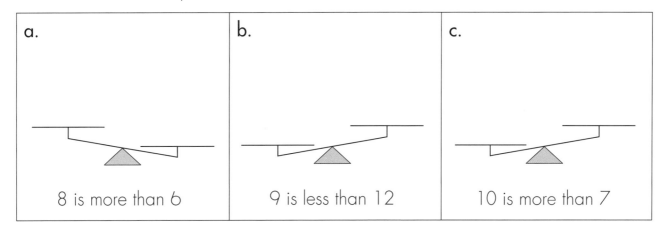

a. 8 is more than 6

b. 9 is less than 12

c. 10 is more than 7

Algebra for All, Yellow Level [Blackline Master | More and Less] 15

Beads and Boxes

Using familiar pictures as unknowns

AIM

Students will investigate the conventions used for representing unknowns. For example, the same variables *must* have the same value and different variables *may* have the same value.

MATERIALS

- Small boxes in 2 different shapes, such as hearts and cylinders (1 of each for each pair of students and for demonstration)
- 1 handful of small flat beads or counters for each pair of students and for demonstration
- 1 copy of the blackline master (opposite) for each student

REFLECTION

Ask, *If △ + △ = 12, what number does each triangle represent?* (6) *How do you know?* (Same shapes must be the same amount.)

1 Show the students the 2 different-shaped boxes and say, *Boxes with the same shape each cost the same amount.* Draw the following on the board.

$$♡ + ♡ = \$8$$

Say, *The total cost for both of these boxes is $8.* Point to the first box and ask, *Can this box cost $3? Why not? How much does it cost?* ($4) *Why?* (Same-shaped boxes must cost the same amount.) Tell the students that each bead represents $1. Call on a volunteer to pick 2 heart-shaped boxes and place 4 beads in each. Discuss how this is the only solution if the unknowns are the same.

2 Draw the following on the board.

$$♡ + ○ = \$6$$

Say, *This time, different boxes can cost the same amount or they can cost different amounts. The total cost of these boxes is $6.* Point to 1 box and ask, *If this box costs $3, how much does the other box cost?* ($3) Give each pair of students 2 different boxes and a handful of small beads or counters. Ask each pair to figure out 2 different ways to make a total of $6. Ask, *If each bead represents $1, how many beads can you put in each box so that the total cost equals $6?* Call on volunteers to share their solutions. On the board, record the answers in a table as partly shown right.

♡	+	○	= 6
3	+	3	= 6
			= 6
			= 6
			= 6

3 Ask the students to complete Question 1a on the blackline master. Discuss all 8 possible answers, including, 0 + 8 = 8, 8 + 0 = 8. If necessary, allow them to use beads to help. Give time for the students to complete the blackline master. Call on volunteers to share their answers.

[Equivalence and Equations]

Beads and Boxes

Name _____

1. Complete the tables.

a.

♥	+	○	= 8
5	+	3	= 8
			= 8
			= 8
			= 8
			= 8
			= 8
			= 8
			= 8
			= 8

b.

♥	+	○	= 11
0	+	11	= 11
			= 11
			= 11
			= 11
			= 11
			= 11
			= 11
			= 11
			= 11
			= 11
			= 11
			= 11

2. Complete these.

a. 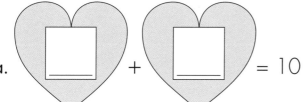 = 10

b. 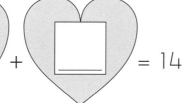 = 14

Shape Maker

Extending repeating patterns

AIM

Students will identify the repeating part in patterns involving geometric shapes. They will also extend repeating patterns in both directions.

MATERIALS

- Pattern blocks for each student and for demonstration
- 1 copy of the blackline master (opposite) for each student

REFLECTION

Discuss how the students identify a repeating pattern. (The same set of shapes keeps repeating.)

1 Seat the students on the floor. Use pattern blocks to show the following pattern.

Say, *Describe this pattern. What is happening to the shape? What will the next shape look like? Why is this a pattern?* Invite several students to share their ideas, then ask for a volunteer to extend the pattern for 3 more pattern blocks. Arrange the students into groups of 3 to copy and extend this pattern for at least 20 pattern blocks. Say, *We call this a "repeating pattern". What part of the pattern is repeating?*

2 Use pattern blocks to show the following pattern.

Repeat the discussion from Step 1. Ask volunteers to extend the pattern in both directions for 3 more pattern blocks. Arrange the students into groups of 3 and have each group copy and extend the pattern for at least 15 pattern blocks in both directions.

3 Ask the students to complete the blackline master. Call on volunteers to share their answers.

[Patterns and Functions]

Shape Maker

Name _____

1. Extend these patterns.

2. Extend these patterns in both directions.

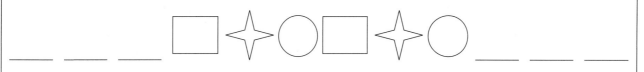

Algebra for All, Yellow Level

Pretty Patterns

Identifying the repeating parts in patterns

AIM

Students will identify the repeating part in patterns involving geometric shapes. They will also translate from one pattern to another.

MATERIALS

- Pattern blocks for each student and for demonstration
- 1 copy of the blackline master (opposite) for each student

REFLECTION

Discuss how the students find the repeating part of a pattern. Encourage them to make non-patterns and to explain why they are not repeating.

1 Seat the students on the floor. Use pattern blocks to show the following pattern.

Say, *Describe this pattern. What will the next block in the pattern look like? Is this a repeating pattern? How do you know? What part is repeating?* Elicit several responses, then ask a volunteer to physically separate the pattern into its repeating parts, as shown below.

Have the students read the pattern aloud, pausing between each repeating part.

2 Show 6 rhombus blocks and ask, *What are these shapes called?* (Rhombuses.) *Can you use these blocks to make a similar pattern to the triangle pattern?* Encourage the students to use the pattern blocks to make a pattern such as the example shown below.

Read the pattern aloud with the students then ask, *What is the repeating part? How is this pattern like the pattern we made with the triangles? Separate your pattern into its repeating parts.* Discuss the similarities.

3 Read Question 1a with the students. Have them draw a loop around the part that repeats before extending the pattern in both directions. After completing Question 1b, ask the students to loop the repeating part in their pattern. Repeat for Questions 2 and 3.

[Patterns and Functions]

Pretty Patterns

Name _____

1. a. Extend this pattern.

   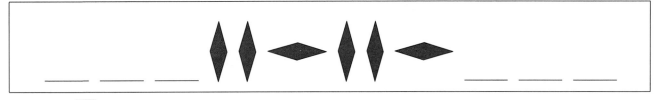

   b. Use ⧖ to make a similar pattern.

2. a. Extend this pattern.

   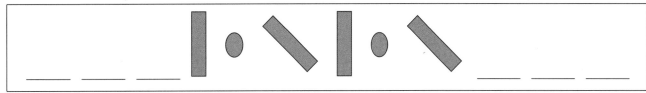

   b. Use ◗■ to make a similar pattern.

3. a. Extend this pattern.

   b. Use ☺ ⬠ to make a similar pattern.

Growing Bigger

Comparing growing and repeating patterns

AIM

Students will identify and extend growing patterns. They will also compare growing patterns to repeating patterns.

MATERIALS

- Pattern blocks for each student and for demonstration
- Identical counters for each student and for demonstration
- 1 copy of the blackline master (opposite) for each student

REFLECTION

Discuss how a growing pattern is different from a repeating pattern. Call on volunteers to use the pattern blocks and the counters to demonstrate the difference.

1 Seat the students on the floor. Use square pattern blocks and counters to make the repeating pattern shown below.

Invite a student to separate the pattern into its repeating parts.

2 Say, *We are going to make this a growing pattern. Each part of the pattern grows by the same amount. Let's make the 2nd part grow by 1 square block and 1 counter.* Show the following 2 parts of the pattern.

Ask, *By how much will the 3rd part grow?* (1 square block and 1 counter.) Invite a student to make the 3rd part. Repeat for 4 more parts. Say, *Describe this pattern. How is it different from a repeating pattern?* Call on volunteers to share their ideas. To demonstrate the difference, compare this pattern to the repeating pattern in Step 1. Ask, *What will the next part of this pattern look like? How does this part differ from the previous part? What shapes do we add each time?*

3 Repeat Step 2 for the following pattern.

Have the students work in pairs to extend the pattern for 4 more parts.

4 Ask the students to complete the blackline master. As time allows, invite students to share their patterns.

[Patterns and Functions]

Growing Bigger

Name _____

1. Draw the next 2 parts in each pattern.

a.

_____ _____

b.

_____ _____

c.

_____ _____

d.

_____ _____

2. Draw 2 different growing patterns using ▲ and ◢.

a.

___ ___ ___ ___ ___ ___

b.

___ ___ ___ ___ ___ ___

Algebra for All, Yellow Level   [Blackline Master | Growing Bigger]   23

Missing Pieces

Identifying missing parts in a repeating pattern

AIM

Students will identify the repeating part in a repeating pattern involving geometric shapes. They will also identify missing parts in the repeating pattern.

MATERIALS

- Pattern blocks
- 1 copy of the blackline master (opposite) for each student

REFLECTION

Ask, *How did you identify the missing parts? Did identifying the repeating part help?* Invite volunteers to share their answers.

1 Seat the students on the floor. Use pattern blocks to show the following pattern.

Say, *Describe this pattern.* Ask questions such as, *What is happening to the shape? What will the next shape in the pattern look like? Is this a repeating pattern or a growing pattern? What part is repeating?* After the discussion, invite a student to extend the pattern for 8 more pattern blocks.

2 Instruct the students to close their eyes. Then remove 2 shapes from the pattern, as shown below.

Have the students look at the pattern again. Ask, *Can you tell me which shapes I have removed? Explain how you know.* Call on a volunteer to replace the missing pieces.

3 Repeat Steps 1 and 2 for the following pattern.

4 Ask the students to complete the blackline master. For each pattern, have the students draw a loop around the part that repeats before drawing the missing shapes. Call on volunteers to share their answers.

[Patterns and Functions]

Missing Pieces

Name _____

Draw the missing shapes.

1.

2.

3.
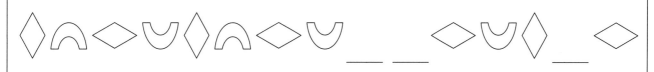

4.

5.
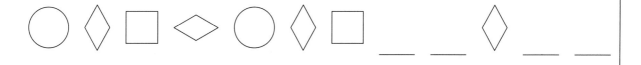

6.

Missing Parts

Identifying missing parts in a growing pattern

AIM

Students will identify missing parts in a growing pattern of geometric shapes.

MATERIALS

- Pattern blocks
- Identical counters
- 1 copy of the blackline master (opposite) for each student

REFLECTION

Refer to the blackline master and ask, *How did you identify the missing parts?* Call on volunteers to share their thinking.

1 Seat the students on the floor. Use pattern blocks to demonstrate the pattern below.

Ask, *Can you describe this pattern? What is happening to the pattern? Who can show me what the next shape will look like? Is this a repeating pattern or a growing pattern? How do you know?* Elicit several responses, then ask a volunteer to extend the pattern for 4 more parts.

2 Instruct the students to close their eyes. Remove 1 part from the pattern, as shown below.

Have the students look at the pattern again. Ask, *Which part have I removed? Explain how you know.* Call on a volunteer to replace the missing part.

3 Repeat the steps for the following pattern.

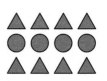

4 Ask the students to complete Questions 1 and 2 on the blackline master. Then ask them to complete Questions 3 and 4 and instruct them to ask a friend to draw the missing parts.

[Patterns and Functions]

Missing Parts

Name _____

1. Draw the next 2 parts of this growing pattern.

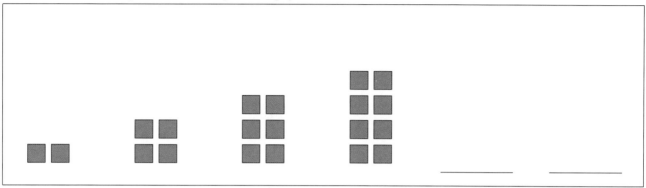

2. Draw the missing parts.

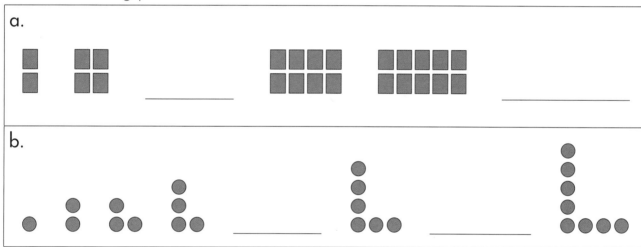

3. Draw 5 parts of a different growing pattern.

4. Draw another growing pattern. Leave 2 parts missing.

Great Grids

Representing repeating patterns as grids

AIM

Students will break repeating patterns into smaller parts to form grid patterns, and then find patterns within the grids.

MATERIALS

- Red and blue connecting cubes for each student
- 3 signs: "4", "/", and "\"
- 1 copy of the blackline master (opposite) for each student

REFLECTION

Challenge the students to find patterns on a calender. They may find vertical patterns such as the days of the week, and diagonal patterns such as counting by sixes (left to right, downward) or counting by eights (right to left, downward).

1 Ask the students to work in groups of 4 to make 10 sets of 3 cubes the same as ▢▢▮. Then have each group join their sets together to make the repeating pattern shown below.

2 Say, *Break your long cube train into shorter trains that are each 4 cubes long.* Tell the students that it is important to keep the shorter trains in order. Demonstrate how to align them vertically as shown right. Position the "4" sign above the demonstration model. Ask, *What pattern are we making? Look at the diagonals. What will the next 4 cubes be? How do you know?* To assist the students to see the pattern, show the "/" and "\" signs. Ask, *Which sign represents this pattern?*

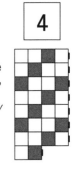

3 Repeat the discussion for breaking the long cube train into shorter pieces that are 6, 8, and then 7 cubes long. Each time, ask the students to predict the pattern the short trains will make. They can shade the grids on the blackline master to check their predictions. Allow students to use the cubes to help them in shading the grids, if necessary.

[Patterns and Functions]

Great Grids

Name _____

1.
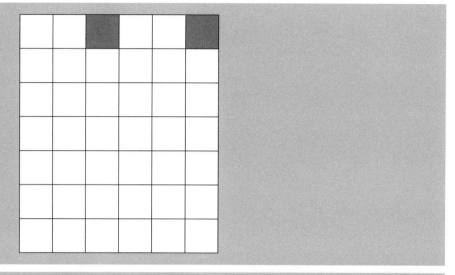

2.
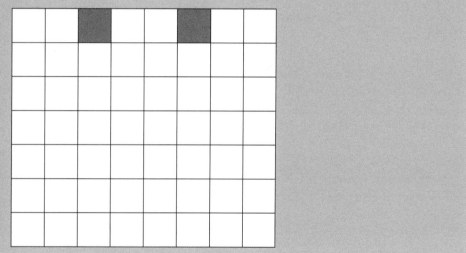

3.
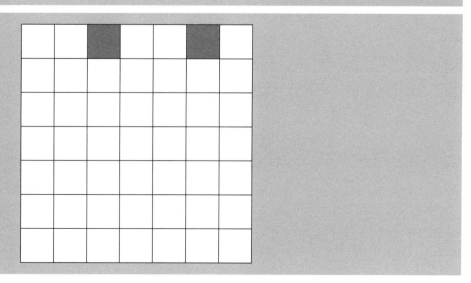

Algebra for All, Yellow Level

[Blackline Master | Great Grids] 29

Odd and Even

Transferring a pattern to a number track

AIM

Students will transfer a repeating pattern such as "red, blue, red, blue" to a number track and identify the parts that are odd and even numbers.

MATERIALS

- Numeral cards for 1 to 20
- Red and blue counters
- 1 copy of the blackline master (opposite) for each student

REFLECTION

Look at the blackline master and read the even numbers aloud with the class. Ask, *Who can see a pattern?* Elicit several responses, then ask, *Is 56 even or odd? Is 81 even or odd? Is 13 odd or even? How do you know? What are some other odd numbers? How do you know?* Discuss the "odd, even" repeating pattern in the number system. Investigate the pattern by starting at different numbers, such as 56.

1 Distribute the numeral cards for 1 to 20. Instruct the students with the cards for 1 to 10 to place them in a line, in order, to make a number track.

2 Distribute 5 red and 5 blue counters, and ask those students to make the repeating pattern "red, blue, red, blue" below the number track.

Say, *Let's say the numbers that are the blue counters.* (2, 4, 6, 8, 10) *What do we call these numbers?* (Even numbers.) *What will be the next even number? Look at the numbers that are red counters. What do we call these numbers?* (Odd numbers.) *Let's say the odd numbers.* (1, 3, 5, 7, 9) *What will be the next odd number?* Point to the numbers in sequence, and read the pattern with the class: *Odd, even, odd, even, odd, even, odd, even, odd, even.*

3 Ask the students with the numeral cards for 11 to 20 to add their cards in order to the number track. Ask, *Is 12 a blue or red counter? Is it odd or even? How do you know?* Ask a volunteer to extend the "red, blue" pattern to 12, and say the even numbers from 2 to 12. Ask, *Is 14 odd or even? How do you know? What are the other even numbers?* Elicit several responses, then call for volunteers to place a blue counter below each of the even numbers and ask them to say the even numbers from 2 to 20.

4 Ask the students to complete the blackline master.

[Patterns and Functions]

Odd and Even

Name _____

Shade the even numbers.

1 2 3 4 5 6
7 8 9 10 11 12 13
14 15 16 17 18 19 20
21 22 23 24 25 26 27
28 29 30 31 32 33 34
35 36 37 38 39 40

Growing Numbers

Translating growing patterns

AIM

Students will translate growing patterns as number patterns.

MATERIALS

- Square pattern blocks
- Numeral cards for 1 to 50
- Blu-Tack
- 1 copy of the blackline master (opposite) for each student

REFLECTION

Refer to the number patterns on the blackline master. For each question ask, *How did you figure out how much each part changed by?*

1 Use the square pattern blocks to make the growing pattern shown below.

Say, *Describe this pattern. Is it a growing pattern or a repeating pattern? Why is it a growing pattern? How many pattern blocks have been added to each part?* Point to each part and ask, *How many pattern blocks are in this part?* Below each part, place the numeral card that matches the number of blocks. Discuss the number pattern. Ask questions such as, *Is it made of odd or even numbers? How much is being added to each part? What will the next part look like? Will the number 15 be in this pattern? Is this a growing pattern or a repeating pattern?* Ensure the students justify their thinking.

2 Repeat Step 1 for the following growing pattern.

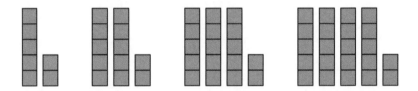

3 Ask the students to complete Question 1 on the blackline master. Call on volunteers to share their answers. Complete Question 2 with the class. Check that students understand that each part must have a similar structure. The pattern below is one possible solution.

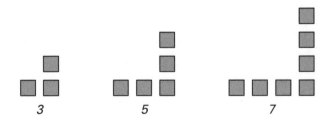

3   5   7

[Patterns and Functions]

Growing Numbers

Name _____

1. Extend these growing patterns. Then write the matching number below each part.

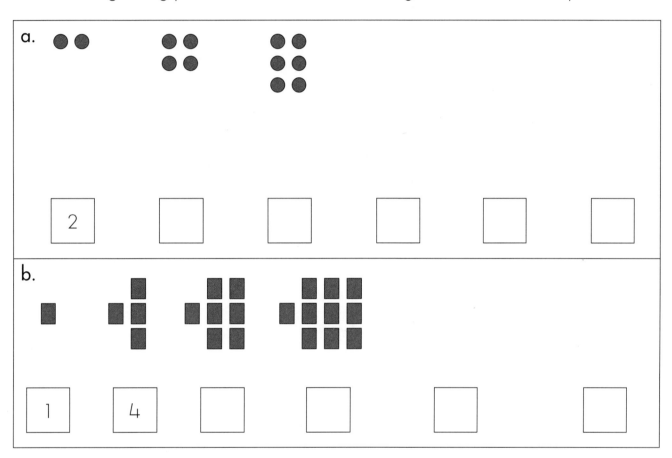

2. Draw shapes to show this number pattern.

| 3 | 5 | 7 | 9 | 11 |

Algebra for All, Yellow Level　　　　　[Blackline Master | Growing Numbers]　33

Nifty Numbers

Building growing patterns with addition and subtraction

AIM

Students will use addition and subtraction to help build growing patterns.

MATERIALS

- 1 copy of the blackline master (opposite) for each student

REFLECTION

Discuss how the students made addition and subtraction patterns. (By continually adding or subtracting the same amount.) Ask, *Can you make an addition or a subtraction pattern where there is also a pattern in the numbers?* (Addition and subtraction for 5.)

1 On the board, write the addition number pattern **2, 4, 6, 8, 10, 12, 14, 16**. Ask questions such as, *What is the 1st number in this pattern?* (2) *What is the 2nd number in this pattern?* (4) *Is the pattern growing or repeating?* (Growing.) *What do we do to the 1st number to make the 2nd number?* (Add 2.) *What do we do to the 2nd number to make the 3rd number?* (Add 2.) Join each pair of numbers and record the amount of change, as shown below.

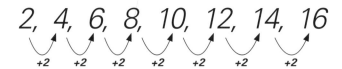

Ask, *Is each number growing by the same amount? By how much?* Call on volunteers to write the next 4 terms.

2 On the board, write the subtraction number pattern **20, 18, 16, 14, 12, 10**. Ask, *Is this an addition number pattern or a subtraction number pattern? How much do we subtract from the 1st number to make the 2nd number?* (2) *How much do we subtract from the 2nd number to make the 3rd number?* (2) Then join each pair of numbers and write **– 2** below each link.

3 Read the blackline master with the class. Then ask the students to complete the questions. Call on volunteers to share their number patterns.

[Patterns and Functions]

Nifty Numbers

Name _____

1. Write the amount that is being added, then complete the pattern.

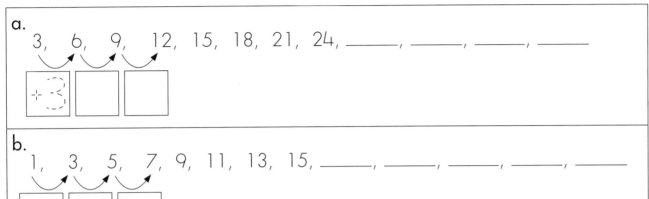

2. Complete these subtraction patterns.

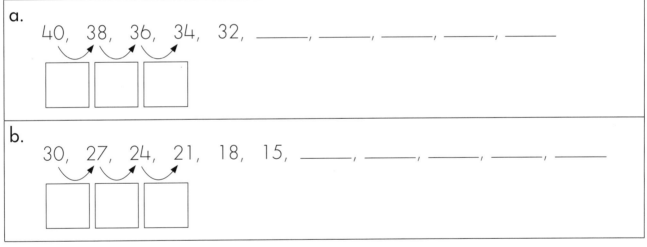

3. Write an addition number pattern. Start at 7.

_____, _____, _____, _____, _____, _____, _____, _____, _____, _____

4. Write a subtraction number pattern. Start at 50.

_____, _____, _____, _____, _____, _____, _____, _____, _____, _____

Fruit Rules

Applying rules to determine output numbers

AIM

Students will apply simple rules from real-world situations and identify the output number when given the input number for a function.

MATERIALS

- Picture cards of apples, oranges, and bananas (at least 15 of each)
- Clear plastic bag
- 1 copy of the blackline master (opposite) for each student

REFLECTION

Discuss how the numbers changed for each special. Make sure the students understand that the rule determines how the starting number will change.

1 Seat the students on the floor and say, *Every time you buy apples, you will receive a bonus of 2 extra apples.* Invite a student to put 3 "apples" into the bag. Ask, *If you buy 3 apples, how many bonus apples will you receive?* (2) *How many apples will you have in total?* (5) *How do you know?* Record the starting and finishing numbers in a table on the board, as shown below.

Apples before	Apples after
3	5

Invite a student to place 5 "apples" in the bag. Ask, *How many apples did Susan buy?* (5) *How many will she have after the bonus?* (7) *How do you know?* Repeat for 12 apples. Ask, *Did the number of apples increase or decrease each time?* (The number increased.) *By how many?* (2)

2 Read the blackline master with the class. Make sure they understand the rules by demonstrating an example of each question. Arrange the students into pairs and ask them to complete the tables. As time allows, call on pairs of volunteers to share their answers.

[Patterns and Functions]

Fruit Rules

Name _____

1. Buy apples and receive a bonus of 2 extra apples.

Rule: Add 2 apples	
Before the bonus	After the bonus
6	8
14	
23	
8	
12	
19	

2. Buy more than 10 bananas and receive a bonus of 3 bananas.

Rule: Add 3 bananas when buying more than 10	
Before the bonus	After the bonus
12	
15	
19	
21	
11	
25	

3. Buy oranges and receive double the number.

Rule: Double the number	
Before the bonus	After the bonus
4	
5	
3	
6	

Algebra for All, Yellow Level   [Blackline Master | Fruit Rules]   37

Helping Hampers

Applying rules to determine input numbers

AIM

Students will apply simple rules from real-world situations and identify the input number when given the output number for a function. They will also identify addition and subtraction as inverse operations.

MATERIALS

- Counters for each pair of students and for demonstration
- Clear plastic bag
- 1 copy of the blackline master (opposite) for each student

REFLECTION

Ask, *When you know the number of items that are bought, how do you figure out how many are taken home?* (Subtract the number of items given to the hampers.) *When you know the number of items that are taken home, how do you figure out how many were bought?* (Add the number of items given to the hampers.) Discuss how addition and subtraction are inverse operations.

1 Say, *The local supermarket asks its customers to donate 2 items for hampers each time we shop. Imagine we buy 15 items. How many will we take home?* (13) *How do you know?* (Subtract 2.) Ask a volunteer to use counters to model the example. Record the numbers in a table on the board, as partly shown below.

Rule: Subtract 2	
Number we buy	**Number we take home**
15	13

Repeat for 12 and 18 items.

2 Say, *Imagine we take home 10 items. How many did we buy?* (12) *How do you know?* (Add 2.) Ask one student to use counters to model the example. Record the numbers in the table. Repeat for 9 items, 15 items, and 4 items.

3 Read the blackline master with the class. Model an example of each question. Ask the students to complete the tables. Call on volunteers to share their answers.

[Patterns and Functions]

Helping Hampers

Name _____

1. The Chan family donates 2 items each time they shop.

Rule: Subtract 2	
Items bought	Items taken home
7	5
15	
18	
	12
	17
	18

2. The Amberley family donates 3 items each time they shop.

Rule: Subtract 3	
Items bought	Items taken home
12	
8	
20	
	8
	20
	15

3. The Warrigal family donates 5 items each time they shop.

Rule: Subtract 5	
Items bought	Items taken home
6	
8	
12	
	6
	8
	12

Guess the Rule

Identifying the rule

AIM

Students will determine simple rules when given input and output numbers for functions. They will also continue to develop the awareness that addition and subtraction are inverse operations.

MATERIALS

- Counters for each student
- 1 copy of the blackline master (opposite) for each student

REFLECTION

Ask, *If the rule is addition, and you know the IN number, how do you figure out the OUT number?* (Add.) *If the rule is addition, and you know the OUT number, how do you figure out the IN number?* (Subtract.) Allow time for discussion and call on several volunteers to share their thinking. Then ask, *If the rule is subtraction, and you know the IN number, how do you figure out the OUT number?* (Subtract.) *If the rule is subtraction, and you know the OUT number, how do you figure out the IN number?* (Add.) Allow time for discussion and invite students to share their thinking.

1 Draw a simple function machine on the board, as shown below.

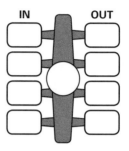

2 Say, *Today we are going to play "Guess the Rule". When we put a number in this machine, it uses a rule to change the number to a new number.* Invite a student to suggest an IN number. Write the number in the 1st IN space on the machine. Apply the rule "add 3" and write the new number in the 1st OUT space. Ask, *Can you guess the rule?* Continue until a student correctly guesses the rule. Then write the rule in the middle of the machine. Repeat for a "subtract 3" rule.

3 Ask the students to complete the blackline master. Direct them to use counters to help them model the IN and OUT numbers. Call on volunteers to share their answers and explain their thinking. Generate discussion by asking questions such as, *Did you add (count on) to the IN number to make the OUT number? Did you subtract (count back) from the OUT number to make the IN number? Did you check that your rule was correct for all the numbers?*

[Patterns and Functions]

Guess the Rule

Name _____

Figure out the rule and write the numbers.

1.

IN	OUT
15	18
25	28
12	15
___	20
16	___
23	___
___	16
___	12

2.

IN	OUT
15	11
18	14
11	7
___	22
10	___
13	___
___	2
___	17

3.

IN	OUT
20	15
8	3
11	6
___	2
25	___
___	27
14	___
___	36

Turnarounds

Exploring the commutative law with addition

AIM

Students will use concrete models to explore addition number sentences and their turnarounds.

MATERIALS

- 1 wire coat hanger
- Red and green clothes pins (pegs)
- 1 copy of the blackline master (opposite) for each student

REFLECTION

Ask, *Is 23 + 49 and 49 + 23 the same or different?* (The same.) *How do you know? Is 45 + 17 and 17 + 45 the same or different?* (The same.) *How do you know?* Call on several volunteers to share their thinking.

1 Show a coat hanger with 6 green clothes pins on one side and 9 red clothes pins on the other, as shown below.

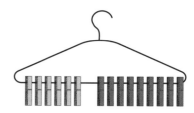

Ask, *How many green clothes pins do we have?* (6) *How many red clothes pins?* (9) *How many clothes pins do we have in total?* (15) *How do we write this as a number sentence?* Write **6 + 9 = 15** on the board.

2 Physically turn or flip the coat hanger, as shown below.

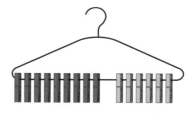

Ask, *What did we do to the coat hanger?* (Turned it around.) *How many red clothes pins do we have?* (9) *How many green clothes pins?* (6) *How many clothes pins in total?* (15) *How do we write this as a number sentence?* Elicit responses, then write **9 + 6 = 15** on the board below the first number sentence.

3 Repeat Steps 1 and 2 for **8 + 9 = 17** and **7 + 4 = 11**.

4 Read the blackline master with the class. Have the students complete the questions. Call on volunteers to share their answers to Question 2.

Turnarounds

Name _____

1. Write 2 matching number sentences for each coat hanger.

a.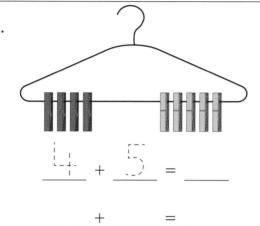

4 + 5 = ___

___ + ___ = ___

b.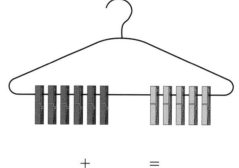

___ + ___ = ___

___ + ___ = ___

c.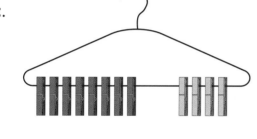

___ + ___ = ___

___ + ___ = ___

d.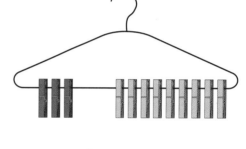

___ + ___ = ___

___ + ___ = ___

2. Draw pictures and write the matching number sentences.

a.

___ + ___ = ___

___ + ___ = ___

b.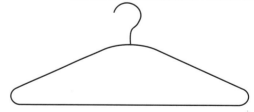

___ + ___ = ___

___ + ___ = ___

Make a Match

Matching turnarounds and writing equivalent number sentences

AIM

Students will match turnarounds and write them as addition number sentences.

MATERIALS

- 6 cards (as shown above right)
- Blu-Tack
- 1 copy of the blackline master (opposite) for each student

REFLECTION

Draw the following card on the board.

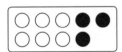

Say, *Look at this card. What will the turnaround card look like?* Invite a volunteer to draw the turnaround card. Then ask, *How do you know?* Discuss the students' answers to Question 3 on the blackline master.

1 Stick the cards to the board in the order shown below.

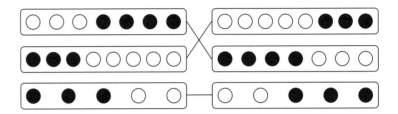

Say, *Look at the total number of dots on each card. Which cards match?* As the students indicate the matching cards, draw lines to join each set, as shown above. Then ask, *How do you know?* Call on several volunteers to share their thinking.

2 Arrange the cards in matched pairs. Call on different students to write 1 number sentence for each pair of matched cards, for example **4 + 3 = 3 + 4**, **3 + 5 = 5 + 3**, and **3 + 2 = 2 + 3**. Then ask, *What do we call these number sentences?* (Turnarounds.)

3 Read the blackline master with the class. Allow time for the students to complete the questions. Call on volunteers to share their answers.

[Properties]

Make a Match

Name _____

1. Draw lines to join a card to its turnaround.

2. Write a turnaround fact to match each card.

a.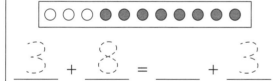

b.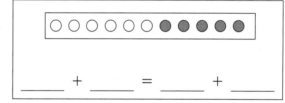

c.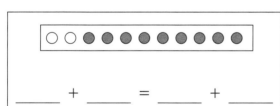

d. ____ + ____ = ____ + ____

3. Draw a pair of turnaround cards and write the matching number sentence.

____ + ____ = ____ + ____

Fish Bowls

Representing addition turnarounds

AIM

Students will use concrete models to represent addition turnarounds for equivalent situations. They will also use the language, "same as".

MATERIALS

- Red and blue magnetic counters (or standard counters and Blu-Tack)
- 1 copy of the blackline master (opposite) for each student

REFLECTION

Discuss the turnarounds for equivalent stories. Say, *Imagine there are 6 cats sleeping in the shade and 5 dogs playing on the grass. How can we turn this story around?* (5 dogs sleeping in the shade and 6 cats playing on the grass.) Call on several volunteers to share their ideas. Ask, *What is the matching turnaround number sentence?* (6 + 5 = 5 + 6)

1 Draw 2 fish bowls on the board. Label each bowl as shown below.

 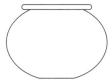

Luke's fish bowl   **Kristy's fish bowl**

Use counters to model the story. Say, *Luke has 3 fish and Kristy has 2 fish. How many fish do they have in total?* (3 + 2 = 5) *Imagine that we swapped the places of the bowls. Will the total number of fish be the same?* (Yes.) *Is 3 plus 2 the same as 2 plus 3?* (Yes.) *What turnaround number sentence can we write for these fish bowls?* Elicit responses and then write **3 + 2 = 2 + 3** on the board. Repeat for **5 + 4 = 4 + 5**.

2 Read the blackline master with the class. Together, complete Question 1a. Ask the students to complete the blackline master. Call on volunteers to share their answers.

[Properties]

Fish Bowls

Name _____

1. Complete 1 number sentence for each pair of fish bowls.

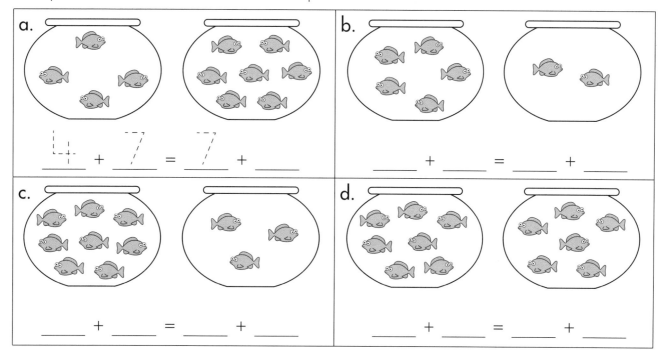

a. __4__ + __7__ = __7__ + ____

b. ____ + ____ = ____ + ____

c. ____ + ____ = ____ + ____

d. ____ + ____ = ____ + ____

2. Draw fish to match the number sentences and then complete the turnarounds.

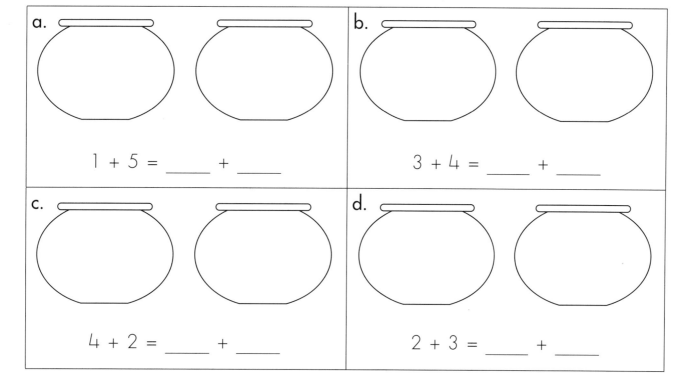

a. 1 + 5 = ____ + ____

b. 3 + 4 = ____ + ____

c. 4 + 2 = ____ + ____

d. 2 + 3 = ____ + ____

Algebra for All, Yellow Level — [Blackline Master | Fish Bowls]

Order Option

Exploring the order for adding more than two numbers

AIM

Students will determine that the order in which numbers are added is irrelevant: The answer remains the same.

MATERIALS

- Overhead projector
- Red, yellow, and blue transparent counters
- Small piece of tagboard or light card
- 1 copy of the blackline master (opposite) for each student

REFLECTION

Write **23 + 34 + 15** on the board. Ask, *If you add 23 and 34 and then add 15, will the answer be the same as when you add 15 and 34 and then add 23?* (Yes.) *How do you know? Is this true for all numbers?* (Yes.) Discuss how, when the same 3 numbers are added in any order, the answer will always be the same.

1 Place 3 red, 4 blue, and 6 yellow transparent counters on the overhead projector in separate groups. Ask, *How many red counters do I have?* (3) *How many blue counters do I have?* (4) *How many yellow counters do I have?* (6) *How many counters do I have in total?* (13) Demonstrate the calculation, as shown below.

Cover the yellow counters with the card. Add the first 2 numbers (3 + 4 = 7) and then uncover the yellow counters and add 7 + 6 = 13. Ask, *How do we write this?* Write **3 + 4 + 6 = 13** on the board.

2 Cover the red counters, as shown right.

Ask, *If we add the blue counters and yellow counters first, and then add the red counters, will the answer be the same?* Call on volunteers to share their ideas. Demonstrate the calculation and write **4 + 6 + 3 = 13** on the board. Say, *"Red add blue add yellow"* is the same as *"blue add yellow add red"*. Cover the blue counters. Repeat the discussion and write **3 + 6 + 4 = 13** on the board.

3 Say, *We used 3 different orders when we added the 3 numbers and the answer was always the same.* Write the 3 different addition orders on the board, as shown below.

3 + 4 + 6 = 13   3 + 4 + 6 = 13   3 + 4 + 6 = 13

4 Read the blackline master with the class. Ask the students to complete the questions. Call on volunteers to share their answers to Question 3.

Order Option

Name _____

1. Write the matching number sentences.

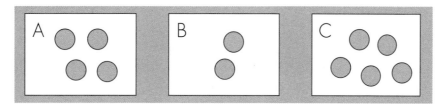

Card A + Card B + Card C   ___ + ___ + ___ = ___

Card B + Card C + Card A   ___ + ___ + ___ = ___

Card A + Card C + Card B   ___ + ___ + ___ = ___

2. Write 3 different ways to find the total.

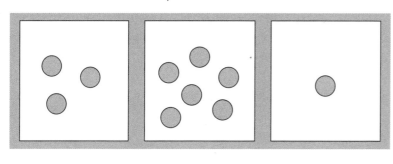

___ + ___ + ___ = ___

___ + ___ + ___ = ___

___ + ___ + ___ = ___

3. Draw some ◯ in each box. Then write 3 different ways to find the total.

___ + ___ + ___ = ___

___ + ___ + ___ = ___

___ + ___ + ___ = ___

Algebra for All, Yellow Level   [Blackline Master | Order Option]

Paper Strips

Investigating the order for adding more than two numbers

AIM

Students will use a length model to further explore adding 3 numbers in different orders.

MATERIALS

- 1 set of paper strips representing the numbers 1 to 10 (use the blackline master opposite)
- Blu-Tack
- 1 copy of the blackline master (opposite) for each student
- Scissors for each student

REFLECTION

Ask, *If we add the same 3 numbers in different orders, will we always get the same answer?* Ensure the students justify their answers.

1 Stick the paper strips for 5, 4, and 6 to the board and label them as shown below.

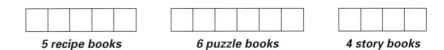

5 recipe books **6 puzzle books** **4 story books**

Say, *Imagine that on a bookshelf there are 5 recipe books, 6 puzzle books, and 4 story books. How many books are there in total?* (15) *Which numbers did you add first?* Call on several volunteers to share their thinking. Then write **5 + 6 + 4 = 15** on the board. Say, *If we add the puzzle books and the story books, then add the recipe books, will the answer be the same?* (Yes.) *How can we write this?* Elicit responses and then write **6 + 4 + 5 = 15** on the board. Ask, *Can we add these 3 numbers another way and still get the same answer?* (Yes.) Invite suggestions, then write **5 + 4 + 6 = 15** on the board.

2 Say, *We used 3 different orders when we added the 3 numbers and the answer was always the same.* Write the 3 different addition orders on the board, as shown below.

5 + 6 + 4 = 15   5 + 6 + 4 = 15   5 + 6 + 4 = 15

3 Give each student a copy of the blackline master and a pair of scissors. After they have cut out the paper strips, ask them to select any 3 strips and write the number of squares on each. Direct them to place the strips in a row and add the 3 numbers in 3 different orders. Ask, *Do you get the same answer each time?* Call on volunteers to share their numbers and addition orders. Repeat for 3 different paper strips.

[Properties]

Paper Strips

Take a Turn

Exploring the order for subtracting whole numbers

AIM

Students will explore subtraction turnarounds and determine that they have different answers.

MATERIALS

- 1 copy of the blackline master (opposite) for each student

REFLECTION

Ask, *Is 23 + 15 + 12 the same as 23 + 12 + 15?* (Yes.) *Is 23 + 15 − 12 the same as 23 + 12 − 15?* (No.) *How do you know? How do we write these as number sentences?* (23 + 15 + 12 = 23 + 12 + 15 and 23 + 15 − 12 ≠ 23 + 12 − 15.)

1 On the board, write **6 + 3 + 5 = ___** and then below this write **6 + 5 + 3 = ___**. Ask, *What do you know about the answers to these number sentences?* Discuss how the answers are the same.

2 Draw a number track from 1 to 20 on the board. Write the sentences on the board, as shown below.

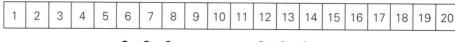

$6 + 2 − 3 =$ ___ $6 + 3 − 2 =$ ___

Ask, *Will the answers be the same for these number sentences?* (No.) Elicit several responses then invite 2 volunteers to write the answers on the board. Have them check their answers by drawing the jumps on the number track, as shown below.

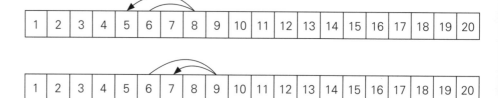

3 Write **6 + 2 − 3 ☐ 6 + 3 − 2** on the board. Ask, *Are these two number sentences equal?* (No.) *What symbol do we write to complete the number sentence?* (≠) Invite a volunteer to write the "not equals" symbol in the number sentence on the board.

4 Draw a number track from 1 to 20 on the board. Use it to complete Question 1 on the blackline master with the class. You may want the students to use a different pencil for Question 1b. Discuss how each number sentence gives different answers. Ask the students to complete the blackline master. Allow time for them to share and justify their answers.

Take a Turn

Name _____

1. a. Draw jumps. Start at 4, add 5, subtract 3.

   | 1 | 2 | 3 | **4** | 5 | 6 | 7 | 8 | 9 | 10 | 11 | 12 | 13 | 14 | 15 | 16 | 17 | 18 | 19 | 20 |

   b. Draw jumps. Start at 4, add 3, subtract 5.

   c. Is 4 + 5 − 3 the same as 4 + 3 − 5? _____

   d. Write = or ≠.   4 + 5 − 3 ____ 4 + 3 − 5

2. a. Draw jumps. Start at 8, add 7, subtract 5.

   | 1 | 2 | 3 | 4 | 5 | 6 | 7 | **8** | 9 | 10 | 11 | 12 | 13 | 14 | 15 | 16 | 17 | 18 | 19 | 20 |

   b. Draw jumps. Start at 8, add 5, subtract 7.

   c. Is 8 + 7 − 5 the same as 8 + 5   7? _____

   d. Write = or ≠.   8 + 7 − 5 ____ 8 + 5 − 7

3. Draw jumps for each. Then, shade true or false.

   | −1 | 0 | 1 | 2 | 3 | 4 | 5 | 6 | 7 | 8 | 9 | 10 | 11 | 12 | 13 | 14 | 15 | 16 | 17 | 18 |

   a. 5 − 4 = 4 − 5   True   False

   b. 10 + 7 = 7 + 10   True   False

Algebra for All, Yellow Level   [Blackline Master | Take a Turn]

No Change

Adding and subtracting zero

AIM

Students will determine that adding and subtracting zero results in no change. They will also explore the idea that when a number is added to zero, the answer is the number added, but when a number is subtracted from zero, the answer is not the number subtracted.

MATERIALS

- Magnetic counters (or standard counters and Blu-Tack)
- 1 copy of the blackline master (opposite) for each student
- Calculator for each student

REFLECTION

Say, *Think of 10 very large numbers and write them down. Use a calculator to add zero to each number. Do they remain the same?* (Yes.) *Now subtract zero from each number. Do they remain unchanged?* (Yes.)

1 Draw an empty piggy bank on the board. Model the story with counters. Say, *Imagine we have no money in our piggy bank. If we saved $8, how much money do we have?* ($8) *How do we write this?* Elicit responses and then write **0 + 8 = 8** on the board. Say, *we have $8 in our piggy bank. What happens if we add zero dollars?* Encourage the students to say the number sentence. Then write **8 + 0 = 8** on the board. Ask, *How is this different from what we did before?* (First, we added 8 to zero then we added zero to 8.) *When we add a number to zero the answer is the number added. When we add zero to a number the answer is the starting number.*

2 Say, *Imagine we have $10 in our piggy bank. If we go to the store but spend no money, how much money do we have left?* ($10) *How do you know?* Encourage the students to say the number sentence. Then write **10 − 0 = 10** on the board. Say, *When we subtract zero from a number, the answer is the starting number.* Repeat for other amounts.

3 Say, *Imagine we have no money in our piggy bank and we want to spend $10. Can we do this?* (No.) *Why not? What if we want to spend $15 or $100, can we do this?* (No.) Write, **When we subtract a number from 0 the answer is the number we subtracted**, on the board. Ask, *Is this statement true?* (No.) *How can we make it true?* (When we subtract a number from zero the answer is *not* the number we subtracted.) Make the change to the sentence and ask, *How can we write this?* Write **0 − 10 ≠ 10** on the board.

4 Ask the students to complete the blackline master. Discuss their responses.

54 [Properties]

No Change

Name _____

1. Put no more money in each piggy bank. Write a matching number sentence.

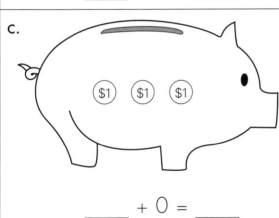

a. 8 + 0 = ____

b. ____ + 0 = ____

c. ____ + 0 = ____

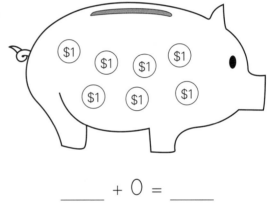

d. ____ + 0 = ____

2. Complete these number sentences.

   a. 6 + 0 = ____   b. 0 + 4 = ____   c. 3 + 0 = ____

   d. 8 − 0 = ____   e. 7 − 0 = ____   f. 2 − 0 = ____

3. Complete these sentences.

   a. When you add 0 to a number _____

   _____

   b. When you subtract 0 from a number _____

   _____

Algebra for All, Yellow Level   [Blackline Master | No Change]

Frogs in a Pond

Making ordered lists

AIM

Students will make an ordered list and record the list in a table format. They will then rewrite the table in a systematic way.

MATERIALS

- 1 copy of the blackline master (opposite) for each student
- 5 identical magnetic counters (or standard counters and Blu-Tack)
- 5 identical counters for each pair of students
- Numeral cards for 0 to 7
- Blu-Tack

REFLECTION

On the board, draw 2 ponds and a table with 8 blank rows and the headings "Frogs in Pond 1" and "Frogs in Pond 2". Show the numeral cards for 0 to 7. Challenge the students to systematically find all the ways to put 7 frogs in the 2 ponds. Demonstrate the 1st step by placing the "zero" card in one pond. Ask, *How many frogs are in the other pond?* (7) Then place the "1" card in the 1st pond and ask, *How many frogs are in the other pond?* (6) Continue until there are 7 frogs in the 1st pond and zero frogs in the 2nd pond. Record the data in the table.

1 Give each student a copy of the blackline master. Draw a pond and the table from Question 1a on the board. Use counters to model the story. Say, *Imagine we have 3 frogs in the pond. How many frogs are out of the pond?* (2) Record the answer in the table. Repeat for 2 frogs in the pond. Ask the students to work in pairs with 5 counters to complete Question 1a.

2 Ask, *How many different ways can 5 frogs be in and out of the pond?* (6) *How can we check that we have all the different ways?* Allow time for the students to share their thinking. Explain that the table could be rewritten systematically to check. Have the students complete Question 1b with their partners, then call on volunteers to share their answers. If the students have not completed the table systematically, draw examples on the board as shown below. Say, *Here are 2 ways that you can write the answers systematically.*

In the pond	Out of the pond
0	5
1	4
2	3
3	2
4	1
5	0

In the pond	Out of the pond
0	5
5	0
1	4
4	1
2	3
3	2

Use counters to demonstrate the 2 ways shown in the tables.

3 Ask the students to complete Question 2. Call on volunteers to share their findings.

Frogs in a Pond

Name _____

1. a. Write the different ways 5 frogs can be in and out of the pond.

In the pond	Out of the pond

b. Write the different ways in order.

In the pond	Out of the pond

2. Chi and Sam wrote all the ways they could place 6 frogs on or off the river bank.

ON	OFF
6	0
2	4
3	3
1	5
0	6
4	2

ON	OFF

a. Write the different ways in order.

b. What did you find? _____

_____

_____

_____

_____

Algebra for All, Yellow Level [Blackline Master | Frogs in a Pond] 57

Mix and Match

Making ordered lists

AIM

Students will make an ordered list and record it in a table format.

MATERIALS

- 1 copy of the blackline master (opposite) for each student
- 5 signs: "blueberry", "chocolate", "strawberry", "nuts", and "jelly beans"
- Blu-Tack

REFLECTION

Ask, *How can we systematically figure out all the different combinations we can make using 5 different ice creams and 4 different toppings?* Elicit several responses. Discuss examples of each method.

1 Give each student a copy of the blackline master. Read Question 1 with the class. Then draw an ice cream cone on the board. Say, *Imagine we can have blueberry, chocolate, or strawberry ice cream, with nuts or jelly beans on top. If we choose blueberry ice cream, what topping can we have?* (Nuts or jelly beans.) *How many different toppings can we have on blueberry ice cream?* (2)

2 Ask the students to complete Question 1. Invite volunteers to share their combinations.

3 Discuss how to make an ordered list. Use the signs to demonstrate the different combinations on the board. Discuss how the students can use the same ice cream and vary the topping, or they can use the same topping and vary the ice cream.

4 Ask the students to complete the blackline master. Call on volunteers to share their answers.

[Representations]

Mix and Match

Name _____

1. Show all the different cones that can be made with these ice creams and toppings.
   - Shade the ice cream: blueberry (blue), chocolate (brown), or strawberry (pink)
   - Draw the topping: nuts or jelly beans

2. Record your cones in the table below.

Ice cream	Topping

Bits in Boxes

Making ordered lists

AIM

Students will make an ordered list and record it systematically in a table format.

MATERIALS

- 8 identical magnetic counters (or standard counters and Blu-Tack)
- 20 identical counters for each pair of students
- 1 small box or container for each pair of students
- 1 copy of the blackline master (opposite) for each student

REFLECTION

Ask the students to explain what is meant by "making an ordered list". Challenge them to make an ordered list for putting 20 counters in and out of the box. Allow them to use counters and a container to help, if necessary. Remind the students that one way is to start with zero counters in the box, then 1, then 2, then 3, and so on. Each time, they have to figure out how many counters are out of the box.

1 Give each student a copy of the blackline master. Draw a box and the blank horizontal table from the blackline master on the board. Place 8 counters on the board and say, *We have 8 counters out of the box. How many are in the box?* (0) Record **8** and **0** in the table. Repeat for 4 counters out of the box.

2 Arrange the students into pairs. Give each pair 8 counters and ask them to complete Question 1a. Encourage them to find all the different combinations in any order that they choose.

3 Ask, *How many ways can 8 counters be in and out of the box?* (9) *How do we know we have all the ways?* Discuss how the best method of checking is to write an ordered list. Then ask the students to complete Question 1b with their partners. Call on pairs to share their lists.

4 Draw the tables on the board, as shown below.

In the box	0	1	2	3	4	5	6	7	8
Out of the box	8	7	6	5	4	3	2	1	0

In the box	0	8	7	1	2	6	3	5	4
Out of the box	8	0	1	7	6	2	5	3	4

Say, *These are 2 ways of recording the list. Are these both ordered? How do you know?* Use counters to demonstrate both ways of making an ordered list.

5 Ask the students to complete the blackline master. Call on volunteers to share their answers.

[Representations]

Bits in Boxes

Name _____

1. a. Write numbers to show different ways that 8 counters can be in and out of a box.

In the box	1	4							
Out of the box	7	4							

b. Write the different ways in order.

In the box									
Out of the box									

2. Figure out how many ways 10 counters can be in and out of the box below. Use counters to help. Record your answers in order in the table.

In the box	Out of the box

Algebra for All, Yellow Level

[Blackline Master | Bits in Boxes]

Reading Pictures

Interpreting pictographs

AIM

Students will interpret vertical pictographs.

MATERIALS

- 1 copy of the blackline master (opposite) for each student

REFLECTION

Refer to the graphs on the blackline master. Ask, *What other questions can we ask about the graphs?* Write the students' suggestions on the board and discuss the answers.

1 On the board, draw the following pictograph.

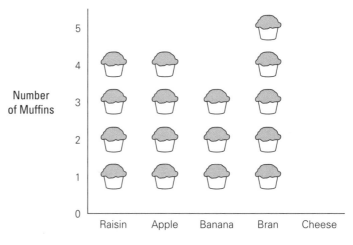

Ask, *What is this graph showing?* (Muffins eaten for breakfast.) *How many types of muffins are there?* (5) *Name the different types.* (Raisin, apple, banana, bran, and cheese.) *How many cheese muffins are there?* (0) *Which muffin is the most popular?* (Bran.) *Which types have the same number of muffins?* (Raisin and apple.) Write the sentences shown below on the board. Call on different students to tell you the missing information.

There are 3 _____ muffins.
There are 2 more _____ muffins than _____ muffins.
There are the same number of _____ muffins as _____ muffins.
There are ___ muffins in total.

2 Read the blackline master with the class. Ask, *How many soccer balls are there?* (2) *How many basketballs?* (4) Ask the students to complete the questions. Call on volunteers to share their answers.

[Representations]

Reading Pictures

Name _____

Look at each graph and complete the sentences below.

1.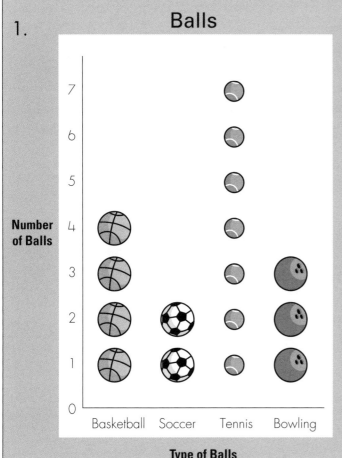

2.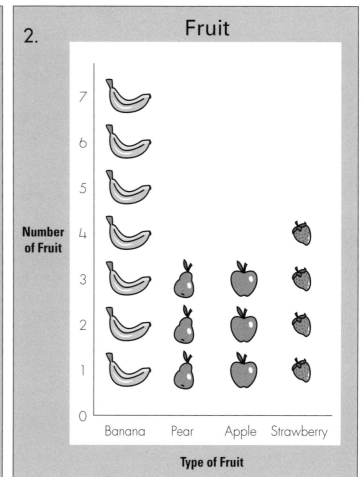

There are _____ soccer balls. There are _____ balls in total. There are 2 more _____ than _____ balls. There are 4 less _____ balls than _____ balls.

There are _____ strawberries. There are _____ pieces of fruit in total. There are 3 less _____ than _____. There is the same number of _____ and _____ .

Peanuts for Grabs

Drawing and interpreting pictographs

AIM

Students will draw and interpret horizontal pictographs.

MATERIALS

- 1 copy of the blackline master (opposite) for each student

REFLECTION

Refer to the blackline master and ask, *What other questions can we ask about the graph?* Write the students' suggestions on the board and discuss the answers.

1 Refer to the blackline master and say, *These students recorded the number of peanuts they grabbed with 1 hand. How many did Hugh grab?* (6) *How many did Lily grab?* (10) *How can we represent the "peanut grab" on the pictograph?* Invite several suggestions. Say, *We can draw graphs either across the page or up the page. This time we will draw across the page How many peanuts will we draw for Hugh?* (7) Direct the students to draw peanuts the first 7 spaces in the 1st row. Repeat for Lily, Chi, Fran, and Mel.

2 Ask, *How can we figure out how many peanuts are grabbed in total?* (7 + 6 + 10 + 12 + 7 = 42) *Who has 2 more peanuts than Chi?* (Fran.) *Who has the greatest number of peanuts?* (Fran.) *How do you know?* Ask the students to complete the question. Discuss ideas such as the fewest number of peanuts has the shortest row; the greatest number has the longest row; and to figure out the total, we add the lengths of the rows.

Peanuts for Grabs

Name _____

1. Some students counted the number of peanuts they could grab with 1 hand.

Hugh	Lily	Chi	Fran	Mel
7 peanuts	6 peanuts	10 peanuts	12 peanuts	7 peanuts

   a. Draw peanuts to show the number each student could grab.

   Peanut Grab

   Students: Hugh, Lily, Chi, Fran, Mel

   Number of peanuts

   _____ can grab 2 more peanuts than _____.

   _____ can grab the fewest number of peanuts.

   _____ can grab the greatest number of peanuts.

   _____ can grab twice as many peanuts as _____.

   _____ can grab the same number of peanuts as _____.

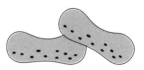

ANSWERS

Equivalence and Equations 1 — Page 7

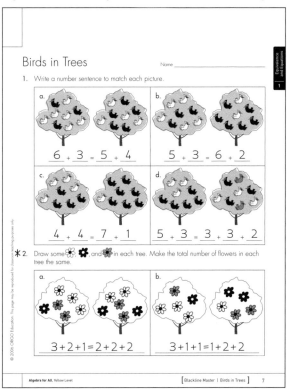

Equivalence and Equations 2 — Page 9

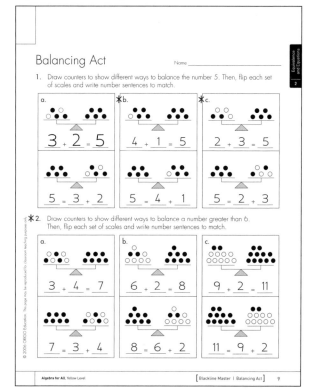

Equivalence and Equations 3 — Page 11

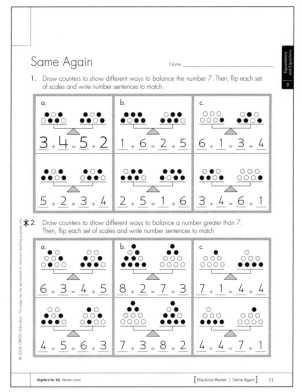

Equivalence and Equations 4 — Page 13

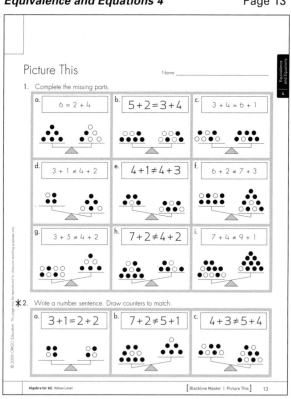

* Answers will vary. This is one example.

Algebra for All, Yellow Level

Equivalence and Equations 5 — Page 15

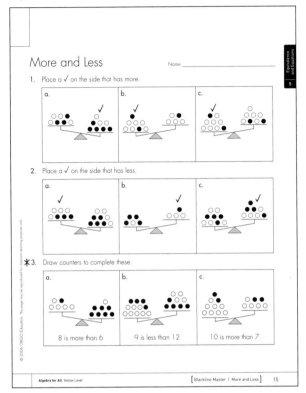

Equivalence and Equations 6 — Page 17

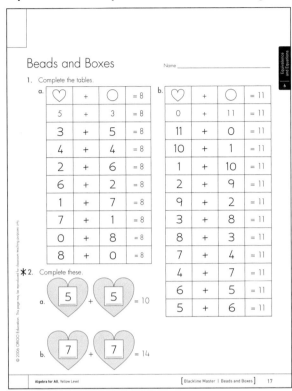

Patterns and Functions 1 — Page 19

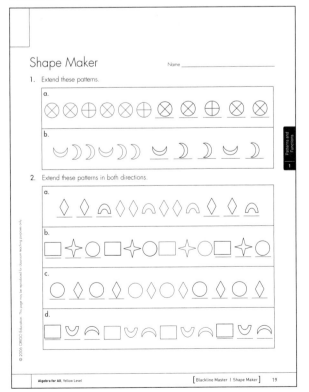

Patterns and Functions 2 — Page 21

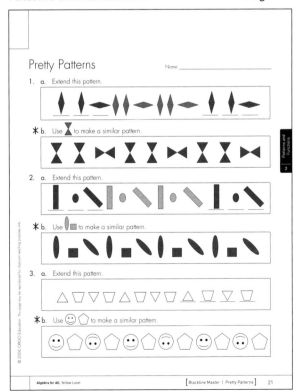

* Answers will vary. This is one example.

Algebra for All, Yellow Level

ANSWERS

Patterns and Functions 3 Page 23

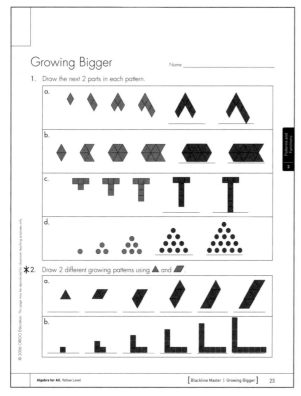

Patterns and Functions 4 Page 25

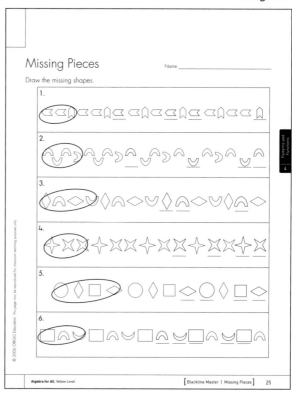

Patterns and Functions 5 Page 27

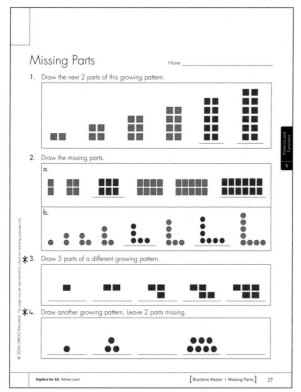

Patterns and Functions 6 Page 29

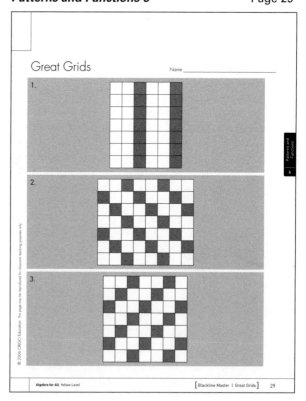

* Answers will vary. This is one example.

Patterns and Functions 7 — Page 31

Patterns and Functions 8 — Page 33

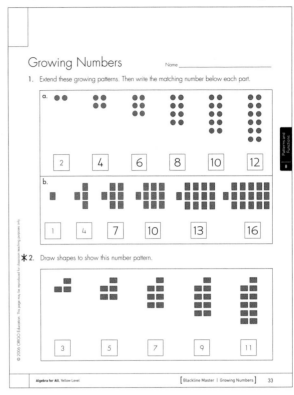

Patterns and Functions 9 — Page 35

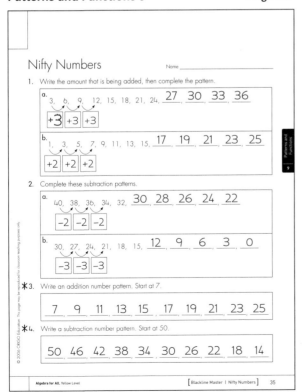

Patterns and Functions 10 — Page 37

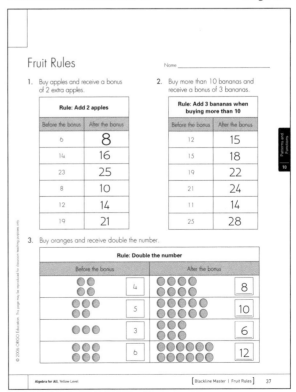

*Answers will vary. This is one example.

Algebra for All, Yellow Level

ANSWERS

Patterns and Functions 11 — Page 39

Patterns and Functions 12 — Page 41

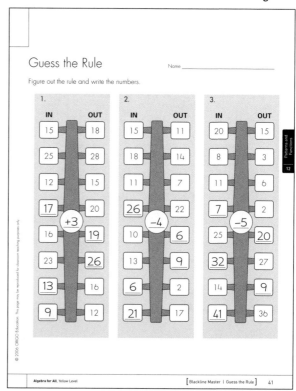

Properties 1 — Page 43

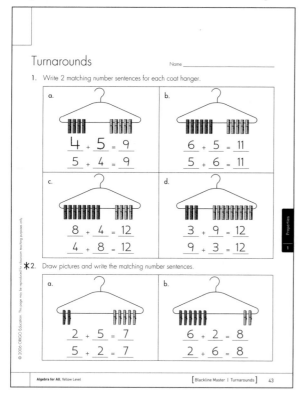

Properties 2 — Page 45

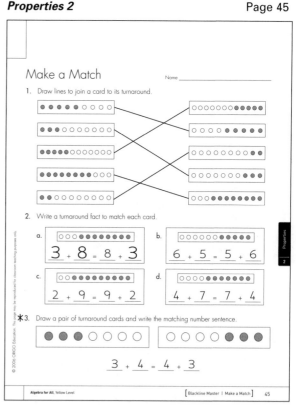

* Answers will vary. This is one example.

Properties 3 — Page 47

Fish Bowls

1. Complete 1 number sentence for each pair of fish bowls.

   a. 4 + 7 = 7 + 4
   b. 5 + 2 = 2 + 5
   c. 8 + 3 = 3 + 8
   d. 7 + 6 = 6 + 7

2. Draw fish to match the number sentences and then complete the turnarounds.

   a. 1 + 5 = 5 + 1
   b. 3 + 4 = 4 + 3
   c. 4 + 2 = 2 + 4
   d. 2 + 3 = 3 + 2

Properties 4 — Page 49

Order Option

1. Write the matching number sentences.

   Card A + Card B + Card C   4 + 2 + 5 = 11
   Card B + Card C + Card A   2 + 5 + 4 = 11
   Card A + Card C + Card B   4 + 5 + 2 = 11

2. Write 3 different ways to find the total.

   3 + 6 + 1 = 10
   6 + 1 + 3 = 10
   3 + 1 + 6 = 10

✱ 3. Draw some ● in each box. Then write 3 different ways to find the total.

   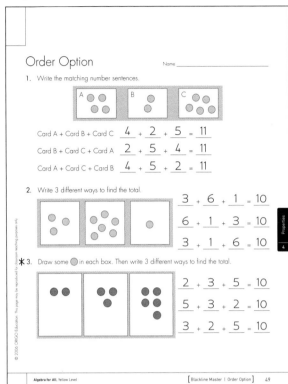

   2 + 3 + 5 = 10
   5 + 3 + 2 = 10
   3 + 2 + 5 = 10

Properties 5 — Page 51

Paper Strips

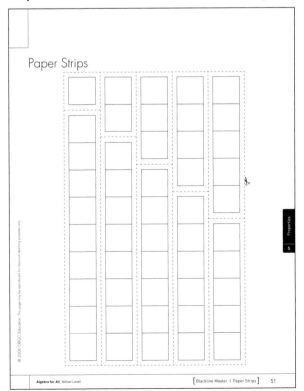

Properties 6 — Page 53

Take a Turn

1. a. Draw jumps. Start at 4, add 5, subtract 3.
   b. Draw jumps. Start at 4, add 3, subtract 5.
   c. Is 4 + 5 − 3 the same as 4 + 3 − 5? No
   d. Write = or ≠.   4 + 5 − 3 ≠ 4 + 3 − 5

2. a. Draw jumps. Start at 8, add 7, subtract 5.
   b. Draw jumps. Start at 8, add 5, subtract 7.
   c. Is 8 + 7 − 5 the same as 8 + 5 − 7? No
   d. Write = or ≠.   8 + 7 − 5 ≠ 8 + 5 − 7

3. Draw jumps for each. Then, shade true or false.

   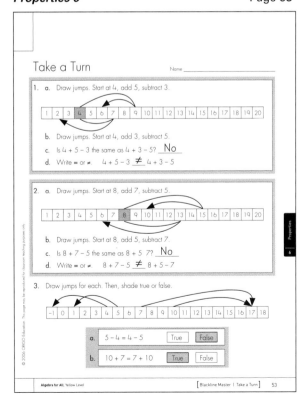

   a. 5 − 4 = 4 − 5   True **False**
   b. 10 + 7 = 7 + 10 **True** False

✱ Answers will vary. This is one example.

ANSWERS

Properties 7 Page 55

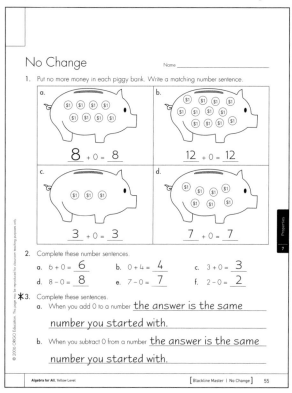

Representations 1 Page 57

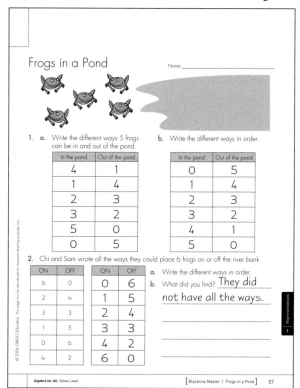

Representations 2 Page 59

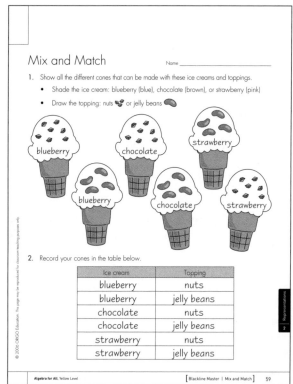

Representations 3 Page 61

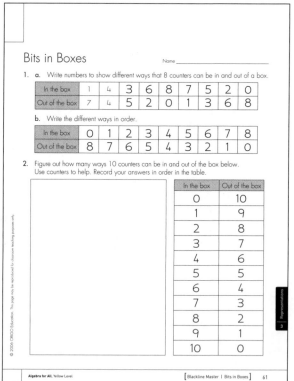

* Answers will vary. This is one example.

Representations 4 Page 61

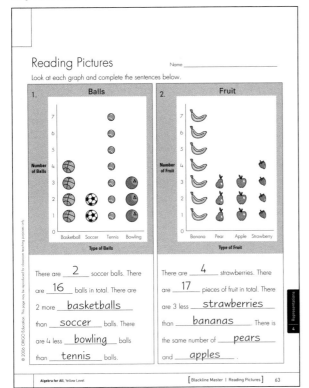

Representations Page 65

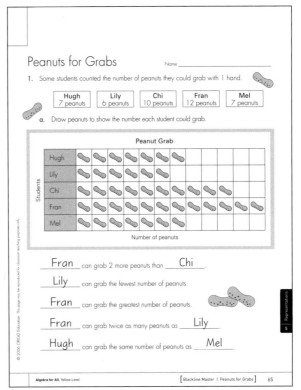

Assessment Summary

Name _____

	Lesson	Page	A	B	C	D	Date
Equivalence and Equations	Birds in Trees	6					
	Balancing Act	8					
	Same Again	10					
	Picture This	12					
	More and Less	14					
	Beads and Boxes	16					
Patterns and Functions	Shape Maker	18					
	Pretty Patterns	20					
	Growing Bigger	22					
	Missing Pieces	24					
	Missing Parts	26					
	Great Grids	28					
	Odd and Even	30					
	Growing Numbers	32					
	Nifty Numbers	34					
	Fruit Rules	36					
	Helping Hampers	38					
	Guess the Rule	40					
Properties	Turnarounds	42					
	Make a Match	44					
	Fish Bowls	46					
	Order Option	48					
	Paper Strips	50					
	Take a Turn	52					
	No Change	54					
Representations	Frogs in a Pond	56					
	Mix and Match	58					
	Bits in Boxes	60					
	Reading Pictures	62					
	Peanuts for Grabs	64					

Algebra for All, Yellow Level